Education
142

热带女神

Goddess of the Tropics

Gunter Pauli

[比] 冈特·鲍利 著

[哥伦] 凯瑟琳娜·巴赫 绘

何家振 译

U0392574

上海远东出版社

丛书编委会

主　　任：田成川

副主任：闫世东　林　玉

委　　员：李原原　祝真旭　曾红鹰　靳增江　史国鹏

　　　　　梁雅丽　孟小红　郑循如　陈　卫　任泽林

　　　　　薛　梅　朱智翔　柳志清　冯　缨　齐晓江

　　　　　朱习文　毕春萍　彭　勇

特别感谢以下热心人士对童书工作的支持：

匡志强　宋小华　解　东　厉　云　李　婧　庞英元

李　阳　梁婧婧　刘　丹　冯家宝　熊彩虹　罗淑怡

旷　婉　王靖雯　廖清州　王怡然　王　征　邵　杰

陈强林　陈　果　罗　佳　闫　艳　谢　露　张修博

陈梦竹　刘　灿　李　丹　郭　雯　戴　虹

目录

Contents

一群蜂鸟的到来让蝎尾蕉一家感到很开心。

　　"如果不是因为你们，可能就没有我，也没有这么多孩子。"一株蝎尾蕉对其中一只蜂鸟说道。

A family of heliconia flowers is enjoying a visit from dozens of hummingbirds.

"I would not exist or have so many children if it were not for all of you,"a heliconia flower says to one of them.

……一群蜂鸟的到来。

... a visit from dozens of hummingbirds.

我的朋友蝙蝠……

My friends, the bats ...

"嗯，我在帮助你们成长为一个大家族。我的朋友蝙蝠也助了你们一臂之力。他们甚至在你们的叶子里建造小帐篷。"蜂鸟回应道。

　　"你有那么多不同的兄弟姐妹来我们这逗留，你们有那么多名字：黑腹蜂鸟、绿紫耳蜂鸟、铜头丽蜂鸟，甚至还有桂红蜂鸟！真让我吃惊。"

"Well, I am helping you to grow a bigger family. My friends, the bats, are also offering a hand. They even build small tents inside your leaves," the hummingbird replies.

"I am amazed by how many of your different brothers and sisters visit us, and what names you have been given: Black-bellied, Green Violet-ear, Coppery-headed Emerald and even a Cinnamon Hummingbird!"

"是的，人们给我们起了一些有趣的名字。我喜欢被叫作火喉蜂鸟。"

"你们是生物多样性和美丽的极好例证。"

"哦，谢谢你。你们蝎尾蕉也很灿烂。"

"Yes, we have been given some interesting names. I like to be called the Fiery-throated Hummingbird."

"You are such a remarkable example of biodiversity, and of beauty."

"Oh, thank you. You heliconia are gorgeous as well."

你们蝎尾蕉也很灿烂。

You heliconia are gorgeous as well.

······最富有想象力的名字······

... the most fantastical names ...

"人们也给我们起了最富有想象力的名字，让人浮想联翩。"

"你们有几百种形状，让人们联想到很多与你们完全不同的事物，比如'龙虾爪'或者'天堂鸟'。"

"People have given us the most fantastical names too, ones that make you imagine all kinds of things."
"Your hundreds of different shapes make people think of something completely different to what you really are. Like 'lobster claws' or 'birds-of-paradise', for instance."

"是这样。但是说到底，我们只是简单的生命体，一种喜欢热带温暖气候的生物。而且，我们喜欢你们的到来。"

　　"你们给我们提供这么多花蜜。不像气候寒冷地区的花儿，你们的花期有4至6周长。"

"That is true. But in the end we are just simple, living creatures, ones that love the warmth of our tropical climate. And we love your visits."

"And you give us such a great supply of nectar. Unlike flowers in cold climates, you blossom for four to six weeks long."

……这么多花蜜。

... such a great supply of nectar.

激励他们成为诗人和艺术家。

... inspire them to become poetic and artistic.

"你知道，我们蝎尾蕉和你们蜂鸟都能让人畅想，激发他们的想象力；我们激励他们成为诗人和艺术家。如今，每个人都如此忙碌，错过了很多美好的事物……"

"我很开心我有这么长的舌头，可以伸进你们的花蕊中，并把花蜜吸上来。"

"You know, we heliconia and you hummingbirds make people dream and use their imagination; we inspire them to become poetic and artistic. Things are missing these days, when everyone is so busy ..."

"I am so happy that I have such a long tongue, one that I can stick deep inside your flowers and suck up the nectar."

"是的，我注意到你的舌头是嘴的两倍长，而你的嘴就已经够长了！"

"你知道吗？在整个动物王国中，我有最大的心脏和大脑。"蜂鸟问道。

"朋友，我看你是在做白日梦了。你可能吃了太多的花蜜……"

"Yes, I noticed that your tongue is twice the length of your beak, which is already quite long!"

"Do you know that I have the biggest heart and brain in all of the animal kingdom?" Hummingbird asks.

"Now you are lost in a world of fantasy, my friend. Maybe you have had too much nectar ..."

······你的舌头是嘴的两倍长······

... your tongue is twice the length of your beak ...

……我拥有最大的心脏。

... I have the biggest heart.

"不不，我的意思是相对于我身体的大小，我拥有最大的心脏。"

"我知道了，一颗小小的心脏——但是相对于你小的身体，却是非常大的。我懂了。现在，你要建议我们不称呼鲸为心脏之王，而是称呼你是吗？"

"噢，不是，我从来不想与鲸竞争。但是，为了采集到足够的食物，我一天要访问大约五千朵花，所以，我必须飞得很快，因此我的确需要一颗强大的心脏。"

"No no, I mean relative to the size of my body, I have the biggest heart."

"I see, a small heart – but very big relative to your tiny body. I get it. Now, are you suggesting that instead of calling the whale the King of Hearts, we should call you that?"

"Oh no, I would never want to compete with the whale. But having to visit some five thousand flowers a day to get enough food, I have to fly very fast and I do need a strong heart."

"我明白了。"蝎尾蕉回应道，"你是一只强大的小鸟。"

"对我而言，你是最华丽的花儿。我想称你为'热带女神'。"

"你说得我脸都红了！你不是坠入爱河了吧？"

"谁不会爱上像你这样的奇迹呢？"

……这仅仅是开始！……

"I see," Heliconia replies. "You are a grand little bird."

"And to me you are the grandest flower. I would like to call you 'The Goddess of the Tropics'."

"Now you make me blush! Are you not perhaps falling in love?"

"Who would not fall in love with a marvel like you?"

... AND IT HAS ONLY JUST BEGUN! ...

……这仅仅是开始！……

Did You Know?

你知道吗?

3m

The heliconia plant can grow up to 3 metres in height. The small true flowers are contained inside the brightly coloured bracts. The bracts block the entrance to bigger birds, providing exclusive access to only a limited number of hummingbirds.

蝎尾蕉属植物可以长到3米高。真正的小花被包含在颜色鲜艳的苞片里。苞片阻止大型鸟类进入,只让有限的几种蜂鸟进入。

Bats use the heliconia leaves for shelter. Some bats even transform the leaves, with a few smart cuts, into a small 'tent' to protect themselves from rain, sun, and predators. Beetles also live in the rolled-up leaves of the heliconia plants, on which they sometimes feed.

蝙蝠把蝎尾蕉的叶子当作庇护所。一些蝙蝠甚至对叶子进行改造,巧妙地切割叶子,将其变成小帐篷,保护蝙蝠不受雨水、阳光和天敌的侵害。甲壳虫也住在蝎尾蕉卷起来的叶子里,它们有时候在叶子上进食。

5 cm

The smallest hummingbird (the Bee Hummingbird) is only 5 cm long, and the largest is the Giant Hummingbird, which is 20 cm long.

最小的蜂鸟（吸蜜蜂鸟）只有 5 厘米长，最大的蜂鸟是巨蜂鸟，有 20 厘米长。

On average hummingbirds flap their wings 50 times per second but some reach up to 200 times, allowing them to fly at a speed of more than 50 kilometres per hour. Hummingbirds gather their food with a hairy tongue that can lick at a rate of 13 licks per second.

蜂鸟平均每秒扇动翅膀 50 次，有些能达到每秒 200 次，使其能够以超过每小时 50 千米的速度飞行。蜂鸟用它们的毛舌舔食食物，它们的毛舌每秒钟能舔 13 次。

Hummingbirds drink flower nectar but will also catch insects for protein. They eat their own weight in food each day. Before migrating over distances of up to 3,000 km, a hummingbird will add 50% of its bodyweight in fat. Hummingbirds can fly up to 20 hours without a break.

蜂鸟喝花蜜，但为了获取蛋白质也抓昆虫。它们每天吃掉与它们自身重量相等的食物。在进行 3 000 千米的长距离迁徙之前，蜂鸟会增加相当于体重 50% 的脂肪。蜂鸟能够连续飞行 20 小时不停歇。

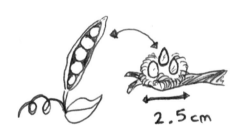

2.5 cm

The size of a hummingbird egg equals the size of a green pea. The nest can be as tiny as 2.5 cm in diameter.

蜂鸟蛋与青豌豆的大小差不多。蜂鸟的巢很小，直径只有 2.5 厘米。

蜂鸟的大脑约占其体重的4.2%，是鸟类大脑中所占比重最大的。蜂鸟很聪明，能记住它们采食过的每一朵花，以及每朵花多久之后能重新长满花蜜。

The hummingbird's brain represents 4.2% of its body weight, proportionately the largest of the bird kingdom. Hummingbirds are very smart and can remember every flower they have visited, and how long it will take a specific flower to refill with nectar.

1200 p/m

When a hummingbird flies, its heart will beat up to 1,200 times per minute; when the bird rests its heart beats 250 times per minute.

当蜂鸟飞行时，它的心脏每分钟跳动高达 1200 次；当蜂鸟休息时，它的心脏每分钟跳动 250 次。

How about all the records the hummingbird holds? All are relative to its tiny size!

蜂鸟保持的各项纪录现在怎样了？所有的纪录都与其微小的体型有关！

Would you be able to remember every flower you have visited and figure out when it would be full of nectar again? Are you able to remember every restaurant you have ever visited and what you ordered at each one?

你能记住你采食过的每一朵花，并算出它们什么时候再长满花蜜吗？你能记住你去过的每一个饭店，在每个饭店各点了什么菜吗？

What do you think of the bat that makes a tent out of a heliconia leaf?

你怎么看蝙蝠用蝎尾蕉做帐篷？

There are some crazy names given to hummingbirds and heliconia. Is there anyone in your family or circle of friends with a crazy name?

蜂鸟和蝎尾蕉都被赋予很多古怪的名字。在你家庭或者朋友圈里，有人拥有古怪的名字吗？

Do It Yourself!

自己动手！

Let's look at the difference between flapping and rotating wings. The hummingbird does not flap its wings but rotates them. The sport of swimming relies on rotating arms. Next time you are in a swimming pool (provided you know how to swim), try the difference between moving your arms up and down, and rotating your arms. You will immediately feel the difference in motion. This first-hand experience will inspire you when you think about engineered solutions. You can, of course, also think about what it would mean if you were able to change the aspect ratio: the width and the length of your arms.

让我们看一下，扇动翅膀和旋转翅膀之间的区别。蜂鸟的翅膀不是扇动，而是旋转。游泳运动靠的是旋转臂膀。下次游泳时（假设你会游泳），试试上下移动手臂和旋转臂膀的差别。你会立即感到运动间的不同。这些第一手经验将会在你做工程方案时给予你灵感。当然，你也可以想一想，如果你能改变你手臂的纵横比，即手臂的宽度和长度的比例，那将意味着什么。

学科知识
Academic Knowledge

生物学	蜂鸟目分为隐蜂鸟亚科和蜂鸟亚科，蜂鸟或者靠固定的几朵花为食，或者住在固定的领地；蜂鸟将花蜜快速地新陈代谢转化为能量；蜂鸟夜间休眠（不活跃）；蜂鸟是社会动物，对它们的幼鸟进行监护；蝎尾蕉是一种重要的植物：授粉共栖；椭圆形叶子。
化 学	花蜜含糖类、氨基酸、脂类、无机离子和生物碱；蝎尾蕉产生的苞液防止食草动物、有害的微藻和真菌的侵害。
物 理	蜂鸟选择特定植物取食源于其喙的特殊形状；蜂鸟通过视觉显示相互沟通；蜂鸟有彩虹色的羽毛；蜂鸟的翅膀可以整周旋转，因此飞行的时候翅膀是旋转而不是扇动；蜂鸟的瞬膜（也被称为第三眼睑）能保护眼睛。
工程学	蜂鸟能够盘旋、向前飞、向后飞、上下飞；灵感来自蜂鸟直升能力的飞行器，性能好于直升机；直升机和微型直升机的区别；飞机的纵横比（机翼的长度和宽度）对飞行很重要。
经济学	蜂鸟是世界上能效最高的飞行者，如果直升机要达到蜂鸟的水平，其能效需要提高27%；共生就如同易货贸易，它没有货币化，没有计入国家统计，然而，它反映了真实的需要和需求。
伦理学	对已经拥有的感到满足，在受到别人赞扬时保持谦卑是一种艺术。
历 史	蜂鸟过去常常因为其羽毛而被猎杀；花鸟画在10世纪的中国出现；蝎尾蕉以赫利孔山的名字命名，赫利孔山是希腊神话中阿波罗、九个艺术和科学女神——缪斯的所在地。
地 理	蝎尾蕉原产于拉丁美洲的热带地区和印度尼西亚的马鲁古省；蜂鸟产于美洲；最小的蜂鸟（吸蜜蜂鸟）产于古巴。
数 学	百分数和分数是相对数；表示相对差的数字是一个标量；百分数可以通过比值乘以100%获得。
生活方式	现在的社会生活方式使我们变得更加孤独，不愿意到朋友或亲人家里串门，更多的人定居在大城市，或生活在国外。
社会学	鸟类和花儿的出现激发人们绘画、写诗和谱曲的热情；中国花鸟画所画的不同物种，象征着不同的品质和情感，比如，竹子象征勇气、谦逊和节节提升，丹顶鹤象征着长寿。
心理学	拥有梦想，可以使人们治愈伤痛、具有创造力和革新精神；弗洛伊德认为在梦想中迷失自我是一种防御机制；白日梦是一种宝贵的资源：化梦想为现实。
系统论	由于栖息地的丧失和气候变化，蜂鸟的生存正受到威胁，其中气候变化影响了它们的迁徙。

情感智慧
Emotional Intelligence

蝎尾蕉

蝎尾蕉赞扬蜂鸟并以谦卑的态度开始了交谈。她分享了她的快乐和惊奇，并且很欣赏蜂鸟的众多名字。蝎尾蕉承认自己的生物多样性，同时也赞赏蜂鸟的美丽。蝎尾蕉保持谦虚，表示她热爱温暖的气候。她非常感谢来访者，清醒地认识到自己和蜂鸟能激发人们艺术和诗歌灵感。她悲叹现代忙碌的生活方式，使人没有时间享受美好的事物。蝎尾蕉从全盘否定转向现实，起初她不相信蜂鸟提供的关于蜂鸟心脏和大脑大小的事实。她想知道蜂鸟是否在自吹自擂，但是当蜂鸟解释这是相对于他身体大小的比例，蝎尾蕉很快理解了。虽然蝎尾蕉在蜂鸟的夸赞下有些害羞，但她还是很自信地问了一个非常个人的问题。

蜂　鸟

蜂鸟承认他在蝎尾蕉繁殖过程中的作用，但是不想揽下所有功劳，所以他指出了蝙蝠所起的作用。蜂鸟对他的种属成员被赋予的各种名字感到高兴，而且喜欢那些奇异的名字。他为被称为美人感到尴尬，但是立即把这个美称送给了蝎尾蕉，并指出蝎尾蕉的生物多样性和众多美好的名字。蜂鸟解释了为什么蝎尾蕉是如此独特，因为她很长的花期为自己提供了更多的食物。蜂鸟希望给蝎尾蕉一个惊喜，并且炫耀他的能力和他的心脏和大脑的比例，但很快表示无意沽取更多名誉。蜂鸟羞怯地表达对蝎尾蕉的爱和欣赏。

艺术
The Arts

鸟和花能给你灵感吗？看一看中国风的花鸟画艺术吧，挑选出你最喜欢的蜂鸟和蝎尾蕉，然后将两者画到一幅彩色油画里。这可能需要很多技巧，但并不需要完美。花点时间研究你所选的两幅图像。想一想你会赋予所选植物和鸟类什么象征意义，你想传达给观看你绘画作品的人什么信息。一旦你确定了背后的寓意，把表达你意图的某些暗示画在图画里。可能起初对外人来说很难看懂，但它可能是其他人如何理解你作品的一个参照。

思维拓展
Systems: Making the Connections

生活中的幸福时常建立在简单的事情上，比如与家人和朋友度过有品质的时光、保持谦逊、专注于生活的品质、摆脱过度的压力等。互惠互利很重要，因为在互利关系中每个成员得到了其他成员创造的好处。这是由共同点很少的物种共同创造的利益，鸟和植物甚至不属于相同的生物王国，但是它们的合作和交流对二者的生存非常重要。人们寻求一致性，在具有相同点的人之间建立起友谊和关系。大自然明确地追求多样性，物种的差异提供了重大甚至惊人的优势。共居比独处更有机会实现没有压力的生活，并有时间专注于生活中特定的重要问题。快节奏的生活让我们很少有时间去欣赏诗歌和艺术，这就减少了我们在幻想和现实之间畅行的机会，花一些时间做做白日梦吧。当社会面临多重挑战时，我们需要更具创造性和革新的精神，同时对那些社会"直觉者"（那些能够不用词汇只用符号进行沟通的人们）保持敏感。我们的观察力要超越那些浮于表面或明显可见的东西。在现代社会忙碌的生活中，我们不仅要调控压力水平，还要建立信心，创造一个能真诚分享我们的感激和赞美的情感空间，这是非常重要的。这为我们提供了信任和适应他人的基础，在爱和关怀得到保障的同时，与他人分享我们的空间，一起开始新的事业。

动手能力
Capacity to Implement

互利是给双方都带来利益的一种关系。想一想你与你的朋友和家庭成员之间的关系，无论在家里，还是在学校，问问自己，你与他们的关系是因为依靠，还是因为彼此间具有真正的互利关系？做一个你认为是依靠关系的清单，无论是你依靠于别人，还是别人依靠你。花些时间仔细想想所列的每一个人。依靠关系会转化为互利关系吗？如果你认为这种转化会发生，那么现实一点，想一想如何执行它，使双方均能受益。注意给予和受益不只限于物质，可能还有非物质价值的交换。就此向你的朋友做个报告，找出建设一个更强大的社区的办法。

故事灵感来自
This Fable Is Inspired by

奚志农
Xi Zhinong

奚志农出生在中国云南大理南部的一个小镇上。他小的时候养过鸭子和麻雀，学生时代，奚志农从科学家、当地住户和公园管理员那里学习大自然的知识。其职业生涯的起点是作为云南省林业局宣传部的一名摄影师。他曾经连续几年拍摄记录滇金丝猴的生活，了解了保护森林对保护滇金丝猴的重要性。在奚志农的职业生涯中，不仅拍摄了濒危物种和受威胁的栖息地，并且记录了大规模非法盗猎活动，给新一代的野生动植物摄影师和摄像师以灵感。他创建的野性中国工作室，是一个促进野生动物摄影和保护事业的非营利组织。2001年奚志农荣获英国野生生物摄影年赛"濒危物种"大奖。

图书在版编目（CIP）数据

冈特生态童书. 第四辑 : 修订版 : 全36册 : 汉英对照 /
（比）冈特·鲍利著；（哥伦）凯瑟琳娜·巴赫绘；
何家振等译. —上海：上海远东出版社，2023
书名原文：Gunter's Fables
ISBN 978-7-5476-1931-5

Ⅰ. ①冈… Ⅱ. ①冈… ②凯… ③何… Ⅲ. ①生态环
境－环境保护－儿童读物—汉、英 Ⅳ. ①X171.1-49

中国国家版本馆CIP数据核字（2023）第120983号
著作权合同登记号图字09-2023-0612号

策　　划　张　蓉
责任编辑　曹　茜
封面设计　魏　来　李　廉

冈特生态童书
热带女神
[比]冈特·鲍利　著
[哥伦]凯瑟琳娜·巴赫　绘

何家振　译

记得要和身边的小朋友分享环保知识哦！
八喜冰淇淋祝你成为环保小使者！